Colt et Wesson the Story of a Rivalry

FAILLE F. Freeman

Copyright © 2021 FAILLE F. Freeman

Tous droits réservés.

ISBN : 9798527869682

TABLE OF CONTENTS

1. Samuel Colt   Pg n°2

2. Daniel Wesson   Pg n°6

3. Colt Paterson   Pg n°10

4. Colt Walker   Pg n°14

5. Colt Dragoon   Pg n°18

6. Smith & Wesson   Pg n°24

7. The civil war   Pg n°30

8. The end of cap & ball   Pg n°36

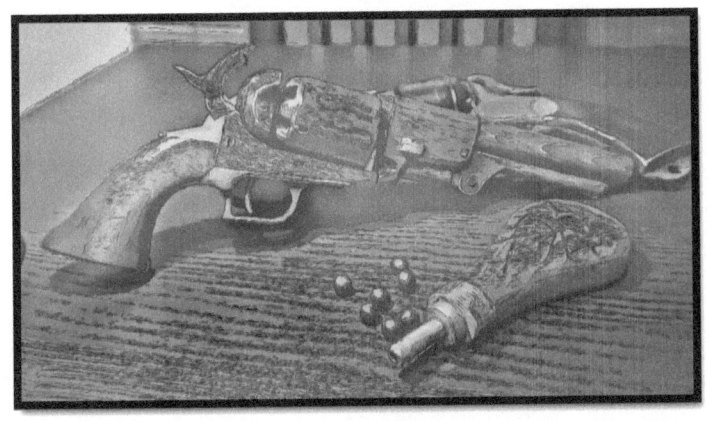

CHAPITRE 1
SAMUEL COLT

Christopher Colt, Samuel's father is a farmer, he becomes widower of his wife Sarah when Samuel is only seven years old. He would convert to a businessman and remarry two years later to Olive Sergeant.

Samuel was born July 19, 1814 in Hartford, Connecticut, he had four brothers and three sisters. Two of these will die during their childhood, one of whom will commit suicide. One of his brothers, John Caldwell, commits a debt murder in 1841 in New York and commits suicide on the day of his execution.

At the age of eleven, Samuel Colt was sent to a farm in Glastonbury to work and attend school. There he read the Compendium of Knowledge, a scientific encyclopedia he prefers to studying the Bible. It contains articles on inventor Robert Fulton and black powder. These are subjects that interest him because he loves explosives and the ideas he draws from them will influence him throughout his life. Reading the Compendium, he learns that Robert Fulton and other inventors "did things that were thought to be impossible, until they came to pass."

In 1829, at the age of fifteen, Colt began

working in his father's textile factory in Ware, Massachusetts, where he was allowed to use the tools and expertise of the workers. He didn't study, but using the knowledge and ideas he had acquired earlier, he made an electric battery that he used to detonate a powder charge on a raft in the waters of Ware Lake. He realized that a properly insulated wire could transmit electricity through water.

Later, following this idea, he will develop waterproof electric cables using tar and will participate in the construction of the telegraph line connecting Manhattan Island to Brooklyn to New Jersey by partnering with Samuel Morse, the inventor of the telegraph. In 1842, on behalf of the Navy, he was even the first to sink on the Potomac, a river in the eastern United States, an old unarmed gunboat named Boxer with an underwater firing mine. electric!

In 1830, he was sixteen and his father sent him to work in the merchant navy in Boston, Massachussetts, to empower him and make him a sailor. On his first trip to Calcutta, India, he heard soldiers talk about the effectiveness of a rifle with two barrels and the inability to create a weapon that could fire five or six times. By observing the helm of the ship, circular and provided with pawls, the concept of the revolver was born in him. Colt takes it into his head that he will be the inventor of this "impossible" weapon.

Colt returned to the United States in 1832, he is now a young man and he returns to work with his father Christopher who finances the production of two weapons, a rifle and a pistol. The guns turn out to be of poor quality because his father, not believing in the project at all, allocates him bad mechanics. The first gun explodes, but the first gun still works correctly.

Then, as he learns, through a chemist working in his father's factory, of the existence of nitrous oxide, Colt begins to travel the roads of the United States and Canada with a portable laboratory, then earning a living by making public demonstrations of the properties of laughing gas, presenting himself as "the famous doctor Coult of New York, London and Calcutta".

A charismatic man, his talents as a speaker are so convincing that he is even asked to heal what appears to be an epidemic of cholera aboard a boat.

It was at this point that he made arrangements to begin manufacturing weapons by hiring gunsmiths from Baltimore, Maryland.

CHAPITRE 2
DANIEL WESSON

Daniel Baird Wesson was born on May 18, 1825 in Worcester, Massachusetts.

Son of Rufus and Betsey Baird Wesson. His father was a farmer and a manufacturer of wooden plows. Daniel worked on the farm all his youth with his father and attended public school until the age of eighteen. He stopped when he apprenticed to his brother and mentor, Edwin Wesson, who was a highly regarded manufacturer of rifles, as well as target pistols, in Northborough, Massachusetts.

His brother Edwin Wesson opened his first store in 1831 in Grafton, Massachusetts, where he specialized in hunting, target and sport percussion rifles. Its rifles quickly became renowned as the best quality percussion rifles in the world, with a quality and finish far superior to any competing product. He moved his store to Northborough, Massachusetts, in 1842, where he remained there until 1848.

During this period, around 1842, Daniel Wesson completed his apprenticeship in his older brother's shop and learned the methods of producing these high quality rifles. Daniel was a fast learner who was always fascinated by mechanics. In 1845, considering that Daniel had acquired sufficient skills, Edwin left him in charge of the shop.

Daniel was married to Cynthia Maria Hawes on May 26, 1847, in Thompson, Connecticut. But his wife's father objected to the couple's engagement fearing he would be a "mere gunsmith" with no future, forcing the couple to flee. They had a daughter and three sons. One of the sons, Frank, died in a train crash on the Vermont Central Railroad. The other two, Walter and Joseph, would later become executives at Smith & Wesson.

CHAPITRE 3
COLT PATERSON

Despite some initial setbacks, Samuel Colt, still motivated by his revolutionary weapon project, is still looking for investors.

At that time, at the beginning of the XIXth century, wars were still practiced with saber and musket. Colt knows his idea is successful. Thanks to his charisma he was able to raise funds, mostly from his family and friends, which amounted to two hundred and thirty thousand dollars at the time, the equivalent of six million dollars today!

With this money, he founded his company, the Patent's Arms Manufacturing of Paterson, in Paterson, New Jersey, then resumed his plans and improved them. His work paid off because on February 25, 1836, at the age of 22, he patented a firearm with a rotating barrel of five shots in .28 caliber.

The first revolver was born, the Colt Paterson!
Its patent will give it the monopoly of the manufacture of revolvers in the United States, United Kingdom and in France.

The Colt Paterson is a black powder revolver with the primers placed on the chimneys behind a barrel with five chambers. The rotation, the alignment and the locking of the barrel are obtained by the manual cocking of the hammer. It is therefore a single action weapon, with a retractable trigger..

The Colt Paterson is loaded from the front of the barrel, barrel removed. To load it, we introduce the necessary charge of powder in each chamber of the barrel, the slightly oversized projectile is then presented to be seated, still from the front of the barrel. To this end, an auxiliary press is used which forcibly inserts the round or conical ball onto the powder charge, ensuring perfect sealing of the charge and its good combustion.

The last step is to lay the primers on the chimneys. The primers composed of mercury fulminate will allow, during percussion, to ignite the powder charges contained in each chamber in order to propel the lead projectile.

The loading procedure, called cap & ball, being slow, it would not have been possible to reload in a combat situation. This is one of the reasons why revolvers usually came with two barrels. But at that time it was still a revolution because with muskets, the best shooters could only load their guns three times per minute.

Le revolver Paterson était fabriqué en quatre

variantes, avec différentes longueurs du canon.

The small Pocket Model N ° 1, in caliber .28, produced in five hundred examples from 1837 to 1840.

The Belt Model N ° 2, with straight butt in caliber .31 and the Belt Model No. 3, with flared butt produced in eight hundred and fifty copies from 1837 to 1840.

The Holster Model N ° 5, said Texas Paterson, in caliber .36.

Word of the invention spreads quickly. Unfortunately, although Colt has an exclusive patent on the guns, he sells few of them and his business struggles to get off the ground.

Still, he ended up selling a few thousand copies of the Holster Model No. 5 to the Texas Navy and its rangers, who were engaged in fierce battles between the Mexicans and the Native Americans three thousand kilometers away.

At that time, Texas was a young republic which had just obtained its independence during the Treaties of Velasco, on May 14, 1836. The Texas Rangers and their captain Samuel Walker lived a permanent war and for them the time to reload a weapon can make a difference in guerrilla-type combat.

CHAPITRE 4
COLT WALKER

Along with his revolver, Samuel Colt also sells long guns, rifled rifles, muskets and smoothbore hunting rifles. Despite everything, his Paterson-based company barely functions and he is on the verge of bankruptcy. The company's shareholders eventually took him over and relegated Colt to the position of sales agent before eventually shutting down operations and selling the company's assets in 1842.

In Mexico, Samuel Walker and his rangers have owned the Colt Paterson for a few years and have gained an advantage in the fighting because where the Mexicans shoot one bullet, the Texas Ranger shoot five. Walker is really very satisfied with this new weapon. He wants and therefore needs more guns.

In 1846, he contacted Colt to provide him with a thousand copies of the revolver, but under certain conditions. Walker wants the new gun to have six shots instead of five. Colt accepts the market but he no longer has a company to produce it and for that he needs one of the best gunsmiths in the country, Edwin Wesson, whose quality long guns he recognizes.

Colt has provided Edwin Wesson with a sample of his new gun and asks them to produce a thousand to start. The Wessons gained fame for their rifle but now, working with Colt, Edwin and Daniel Wesson entered the revolver business. In January 1847, Captain Walker also ordered one of their .44 caliber rifles from them and promised them a one-thousand-piece contract that would never materialize.

With the help of the Wessons, Colt will offer a new .44 caliber revolver: the Colt Walker, named after its sponsor.

The Colt Walker Model 1847 has all the characteristics of the revolvers of the following years, that is to say few moving parts, trigger guard cast with the guard, lever-ram, nose of the hammer comprising a notch intended for the sighting, barrel with six bedrooms. It is a 2kg empty monster, with a barrel of almost 23cm, always in simple action, possessing an elongated barrel and firing ogival .44 caliber bullets propelled by 3.25g of black powder. It is the biggest, heaviest and most powerful black powder revolver ever made, with a nine inch barrel, its firepower would not be matched until eighty ten years later with the modern .357 magnum. Its practical engagement distance is over ninety meters.

It loads in the same way as the Colt Paterson,

except that the ramming lever is now present on the weapon and it is no longer necessary to disassemble the barrel to load the barrel. The loading step, cap & ball, still tedious in the field but no longer impossible, the appearance of paper cartridges makes its debut.

These paper cartridges consist of an extremely flammable paper tube that already contains the powder charge inside with the "bullet," the round or ogival bullet, at its end. In other words, the black powder is already weighed, inserted into a container, and topped with the bullet. The ancestor of the cartridge.

CHAPITRE 5
COLT DRAGOON

In 1848, Samuel Colt released the Colt Dragoon Model 1848, named after the infantry corps that ordered it.

The US Army adopts it as the successor to the Colt Walker, it is of the same caliber, the .44 caliber, but two hundred grams lighter, with a shorter barrel and a seven and a half inch (19cm) barrel. It uses the characteristics of most Colt muzzle-loading revolvers. There are three variations.

The first model has, unlike the Colt Walker, a rammer blocked at the end of the barrel. The main V-spring operates directly on the hammer. The notches of the barrel are round. The trigger guard has a square back. About seven thousand copies were produced in 1848 and 1849.

The second model had a V-spring at the start of production, which was later replaced by a straight spring. Between the spring and the hammer is an intermediate roller, mounted on the hammer. The notches of the barrel are rectangular. About two thousand seven hundred copies were produced between 1849 and 1850.

The third model has a round back trigger guard. A few weapons of late manufacture were equipped with an eight-inch barrel. Some had a removable butt. Nine thousand four hundred copies produced between 1851 and 1861.

The Dragoon is a weapon for mounted troops, it is not carried in a holster on the belt but in a holster in front of the saddle. About half of the Colt Dragoons were purchased by the US government to equip the cavalry, the rest were for the civilian market, and seven hundred were sold in England between 1852 and 1853.

It was an immediate success and Colt's name became synonymous with revolver. More than nineteen thousand pieces will be sold between 1848 and 1861

With his business now successful, Colt is looking to expand its operations. In 1851, he became the first American manufacturer to open a factory in England. He plans to buy a property along the banks of the Connecticut River in Hartford to build a new factory there. He bought the land at a very low price due to its vulnerability to flooding and had a three-kilometer dike built on it to protect his operations from flooding from the river.

The new factory, the largest privately owned manufacturing facility in the world, opened in 1855 under the new company name, Colt Patent Fire-Arms Manufacturing Company. Colt had a distinctive dome installed above his factory to draw more attention to his business.

He sells his products through a variety of different outlets. It employed a small team of traveling salespeople and also used merchants who acted as wholesale agents capable of selling products to retailers in large quantities. Colt also accepts orders direct from wealthy and powerful customers looking for custom parts. He marketed his product through advertising campaigns focused on romantic displays of Western heroism, and also by designing rich ornamental pieces for display at international fairs. These strategies have proven to be very successful. But Colt's groundbreaking idea lies above all in industrializing the manufacturing of its weapons by manufacturing the world's first assembly line.

The same year, while Daniel Wesson was intensely occupied with a percussion rifle, his brother Edwin created a new revolver, based on an 1837 patent granted to his friend Daniel Leavitt. After developing the first model of the Leavitt, Edwin improved it by giving it a six-shot barrel (instead of three originally, with a hand-rotating barrel) and special mechanics with the help of Stevens and Miller, two other gunsmiths. This revolver is today known under the name of Wesson, Stevens & Miller Dragoon Revolver, it was built in very few copies in May 1848.

Edwin will continue to improve this weapon and will file in 1848 a patent for a revolutionary mechanism of rotation of the barrel, consisting of two toothed gears engaged one in the other. It is a six-shot .31 caliber revolver. With its rounded butt, it closely resembles the single-shot muzzle-loading pistols still very common at this time. Just like these pistols, the hammer of this revolver is mounted on the outer right side.

On January 29, 1849, 38-year-old Edwin Wesson died unexpectedly of a heart attack.

CHAPITRE 6
SMITH & WESSON

Daniel has lost a brother, a mentor and a partner and his death will bring down the company. The Wesson Rifle Company, and all its machinery, went on public sale on November 22, 1849 despite all efforts by the Wesson family to prevent it.

Since the sale of the Wesson Rifle Company and thus the loss of its biggest competitor, Colt is more powerful than ever because he controls the production of weapons in the United States.

In 1850 he manufactured the Colt 1851 Navy, using the same technology as his previous weapons but in caliber .36. Lightened by about six hundred grams, it weighs only 1.2kg. It is also more maneuverable, much appreciated for its handling and precision, like Wild Bill Hickock.

Despite its name, the 1851 was used mostly by infantry, with soldiers appreciating its maneuverability and low recoil. The .36 caliber was, however, intended for use by sailors because the .44 Colt Dragoon were too heavy to be worn on the belt. About thirty-five thousand Colt Navy were purchased by the US government, including twenty thousand for the army and fifteen thousand for the navy. It is a weapon that sold over one hundred and fifteen thousand copies until 1873.

For his part, Daniel Wesson will then meet a man who will seal his destiny by giving birth to a mythical name.

Horace Smith was employed by the United States Armory Service from 1824 to 1842, when he moved to Newtown, Connecticut. During the 1840s, he was employed by various gunsmiths and then settled in Norwich, Connecticut. He then appears as a partner of Cranston & Smith. We know that during his stay in Norwich, he launched into the manufacture of guns for whaling and some even attribute to him the invention of the shell to kill whales.

With the help of Horace Smith and Daniel Leavitt as shareholder, Daniel Wesson bought the company with all his tools on March 5, 1850. The company was called the Massachusetts Arms Company, an outgrowth of the Wesson Rifle Company and established itself in Chicopee Falls. , Massachusetts. Its purpose was to produce weapons, including the Wesson & Leavitt revolver developed by Edwin.

They were produced from 1850 to 1851. Although precise and ingenious, Edwin Wesson's system is afflicted with crippling flaws which will help spell the end of this weapon already outdated at the time of its appearance. Without taking into account the rapid wear and the easy fouling of the pinions, the system presents a clear disadvantage because it requires the dismantling of the barrel for the loading and the placement of the primers on the chimneys. It was rejected by the army and barely eight hundred and fifty pieces were manufactured.

Samuel Colt, angry with Stevens and Miller because they had worked for a competitor while they were still employed with him, was the cause of the delay in patenting this revolver and he continued to sue for trifles, until he won a patent infringement lawsuit on June 30, 1851. This lawsuit was resounding and the company Smith & Wesson then had to limit its production of revolvers to relatively unpopular designs until 1857, date on which Colt's patent will expire.

In 1852, Horace Smith and Daniel B. Wesson joined forces in order to develop the weapons store which would be named Smith & Wesson.

In 1854, they founded in Horace Smith's workshop in Norwich, Connecticut, the Smith & Wesson Company, in association with entrepreneur Cortlandt Palmer, to develop magazine firearms, the Volcanic rifle, the first rifle to manual repeat, and a lever pistol. Horace

Smith developed a new cartridge, the Volcanic, which he patented in 1854.

The Smith & Wesson Company was quickly renamed Volcanic Repeating Arms Company in 1855 and had new investors including Oliver Winchester, a shirt maker.

The Volcanic Repeating Arms Company obtained all rights to the Volcanic models (rifle and pistol versions were already in production at that time) as well as ammunition, from the Smith & Wesson Company. Wesson remained as plant manager for eight months before joining Smith in founding the Smith & Wesson Revolver Company after being patented by Rollin White a former Colt employee, for his rear-loading barrel for paper cartridge.

In 1856, Horace Smith and Daniel Wesson left the Volcanic Repeating Arms Company to found a new company and manufacture a newly designed revolver-cartridge combination.

The Volcanic Repeating Arms Company is therefore reorganized for the first time under the name of New Haven Arms Company in 1857, it continues to produce Volcanic rifles and pistols. In April 1857 Benjamin T. Henry was hired as director of the factory.

Also in 1857, the patent on the manufacture of barrel revolvers, issued by Colt in 1836, fell into the public domain and Smith and Wesson jumped at the opportunity. They therefore began to produce a revolver designed to fire rimfire cartridges.

The combination of the metal cartridge and the rear loading of the barrel will allow them to design the Smith & Wesson model n ° 1 of which they will be the sole manufacturers for more than ten years. This model, designed for the .22 caliber rimfire cartridge, has met with great commercial success in the civilian sector. More than two fifty thousand copies will be sold between 1857 and 1882, in three different versions.

Benjamin T. Henry begins to experiment with the new rimfire cartridges and modifies the Volcanic sub-guard lever to adapt it to the new cartridges. In 1861, Henry will develop the loading by the breech, by integrating a magazine of fifteen cartridges. He thus created the first barrel-less repeating rifle, the famous Henry rifle.

.

CHAPITRE 7
THE CIVIL WAR

After Abraham Lincoln won the presidential election of November 1860 through an anti-slavery stance, the first seven slave states declared their secession from the country to form Confederation. War broke out in April 1861 when secessionist forces attacked Fort Sumter in South Carolina, just over a month after Lincoln's inauguration.

Colt based his fortune on the Mexican-American War, and a new war would benefit him. It supplies the army with its weapons, but this army is soon on the verge of splitting in two by waging a war against itself.

Samuel Colt then sells in both camps. In 1860 he sold arms to the southern states for the current equivalent of three million dollars. It's a risky decision for a northern businessman. The press does not fail to treat him as a traitor by defaming him in the public square. His reputation is destroyed!

Smith and Wesson were more careful and didn't sell anything south.

When war is declared Samuel Colt stops selling arms to the southern states. In 1960, he was forty-six years old and was appointed colonel, for Connecticut. He had to go to the front with a regiment of four hundred men, but he wanted to choose them himself. The governor, annoyed by Colt, will demand the dissolution of his regiment. Colt will never participate in the war.

After various attempts to lighten the Colt 1848 Dragoon, the Colt firm decided to develop a .44 caliber revolver using the carcass of the Colt 1851 Navy: the Colt 1860 Army, initially called New Model Holster Pistol.

The use of stronger steel made it possible to manufacture .44 caliber guns, thirty percent lighter than those of the Dragoon. The Colt 1860 Army was the successor of the Colt 1848 Dragoon without replacing it, both models were manufactured until 1873. The Colt 1860 Army was the weapon of the mounted troop and the infantry, it was either carried in a holster in front of the rider's saddle or in the holster of an officer or infantryman, on the belt.

In 1861, the US Navy ordered seven hundred and fifty Colt 1860 Army, the first variant with a fluted barrel. Not being reliable, because the barrel tended to explode, these revolvers were replaced by Colt 1861 Navy the same year.

By April 1862, the Colt factory had already delivered thirty thousand of these revolvers to the government for the price of twenty-five dollars each. In total, more than one hundred and twenty-seven thousand Colt 1860 Army were delivered to the Northerners during the Civil War. The remainder of the production was destined for the civilian market, although many of these weapons were purchased to arm private militias, or by the military as a complementary weapon.

The carcass or plate of the Colt 1860 Army corresponded to that of the Colt 1851 Navy, adapted to the larger diameter of the barrel in caliber .44. The stock was longer and the barrel was round, with slimmer lines than the Dragoon's. The ramming lever was a rack system.

More than two hundred thousand copies will be produced between 1860 and 1873.

In 1861, with the Civil War, Smith & Wesson adapted their Model n ° 1, bringing it to a caliber .32, which gave the Smith & Wesson Model n ° 2.

The Smith & Wesson Model 2 Army is a single-action rocking revolver, on which the barrel swings upwards, articulated on the front end. It can be identified by its octagonal barrel, smooth cylinder without flutes and the flat shape of the grip stock. Revolvers were produced with five or six inch barrels. Seventy-seven thousand copies were made between 1861 and 1874.

For a Civil War soldier, owning a revolver as a back-up weapon was important, so Smith & Wesson's Model # 1 and # 2 revolvers became popular demand with the outbreak of war. American civilian. Soldiers and officers on both sides of the conflict have made private purchases of guns to defend themselves. Lieutenant-Colonel George Armstrong Custer is known to have owned a pair of engraved S&W Army Model No. 2 revolvers.

As the Civil War ended, demand for the S&W Army Model # 2 declined and Smith & Wesson focused on developing weapons in heavier calibers suitable for use at the US border.

The Civil War did not succeed for the New Haven Arms Company because it was unable to assert its weapons with the army. However, she did manage to sell one thousand seven hundred and thirty Henry rifles to the United States government, plus a few hundred directly to combatants. Rather, this situation will prove to be a godsend as it forces him to turn to other markets, such as exports and sales to individuals, which will remain good outlets when the war is over.

In 1962, Colt tried to redeem himself a reputation, he worked to exhaustion and fell ill.

He died of gout on January 10, 1862 at the age of 47. Leaving a fortune of fifteen million dollars at the time, the equivalent of three hundred and fifty million dollars today. He leaves the management of his business to his pregnant wife Elizabeth in the middle of the war.

CHAPITRE 8
THE END OF
CAP & BALL

At the time of his death in 1862, Colt was one of the richest men in the country. He produced over four hundred thousand firearms in his lifetime and made his brand the best known ever. His wife, Elizabeth, took over the business and led it through one of the most difficult times in its history.

On February 5, 1864, about an hour after the start of their working day, workers noticed fire and smoke rising from the attic of the Colt factory. A steam gong alarm sounded and workers rushed to the attic with a hose, but no water came out. The fire quickly spread throughout the facility, burning it down to the foundations. Authorities have never determined the cause of the blaze, although unconfirmed rumors of Confederate arsonists have become a popular explanation among local residents.

The burnt-out Colt's Manufacturing Company plant in Hartford remained closed for two years. Elizabeth will play a key role in rebuilding the arsenal.

Faced with two million dollars in damage and a crumbling facility, Elizabeth Colt had a decision to make, take on the daunting task of rebuilding on site or moving the business to a more pristine or fully operational location. Although her husband didn't feel the need to insure his buildings, Elizabeth Colt did, and she luckily took out insurance on Colt-made buildings shortly after Samuel died. She used that insurance money to rebuild the plant at her existing site in Hartford.

In 1866, after the war, Oliver Winchester took full control of the New Haven Arms Company and reorganized it once again as the Winchester Repeating Arms Company.

He developed the Henry rifle to make it the first Winchester rifle, the famous Winchester 1866 model nicknamed Yellow Boy. This model, like the Henry, used rimfire ammunition and an efficient loading magazine. Huge success with over one hundred and eighteen thousand pieces sold between 1866 and 1873. And so the name Winchester became synonymous with repeating rifles, the weapon that conquered the West.

The Winchesters are characterized by their sub-guard cocking lever, trigger height mechanism, which allows the empty cartridge to be ejected quickly and at the same time to load a new cartridge into the chamber by a back-and-forth movement. -Comes from the lever, operated by the shooter. This principle of operation, simple and ambidextrous, allows a higher rate of fire than that of bolted weapons and allows to keep the target in its line of sight during rearming.

In 1869, the Smith & Wesson company developed a large, single-action, frame-break revolver with an automatic empty case ejector, and first produced in 1870, in the .44 S&W American and .44 Henry calibers.

The design is known as Smith & Wesson Model # 3. It recharges in less than thirty seconds and is therefore faster than a Colt.

Its carcass is reinforced to bring the caliber from .38 to .44. This model is also equipped with a break and an automatic ejection which accelerates and further facilitates the reloading. It is proposed to the Military Commission responsible for the approval of armaments with the regulatory caliber .44, and a 15 gram bullet pushed by a charge of 25 grains of black powder. After placing a first order of one hundred copies, the American Army ordered 28,000 between 1870 and 1874.

Before the war Smith and Wesson had sold a little over eleven thousand revolvers, between the years 1860 and 1868 they will sell more than one hundred thousand!

In 1871, the Russian military attaché in Washington, General Alexandre Gorloff contacted the firm to discuss a large order for one hundred and thirty-one thousand copies to equip the Russian officers. The model is modified a bit, especially with the Russian war caliber very close, but different from the American .44. In 1873, the Smith & Wesson Model n ° 3 said Schofield changed to caliber .45.

Elizabeth Colt got the factory back on track and continued to innovate until she produced the most famous Colt in history: The Colt Single Action Army (SAA) aka Colt Peace Maker (the Peace Maker).

It was developed for the United States Cavalry in 1872 and adopted by the United States Army in 1873 in the .45 Long Colt caliber. It was perhaps the most widely used weapon in the American Wild West, costing at the time thirteen dollars apiece, the price charged to the army, in the civilian market, the price was ten dollars in 1875 and sixteen dollars in 1897. This weapon was manufactured until 1941, with a total of almost thirty-six thousand units produced in about thirty calibers.

A weapon so effective and formidable at the time that it has been said: "God created men, Samuel Colt made them equal" or "Abraham Lincoln made men free, Samuel Colt made them equal".

Excellent design and very powerful for the time, this weapon obtained a test of the Service of the Material of the American army at the end of 1872 like its direct competitor, the Smith and Wesson N ° 3, both in caliber. 44 American or .44 Russian.

As the .44 caliber cartridges were deemed insufficient, Colt developed the .45 Long Colt cartridge, initially with a 255 grain (16.5g) projectile and a 40 grain (2.59g) charge of black powder, that is, is this ammunition which was adopted as an ordinance cartridge, with a charge of thirty grains.

After further testing, requested by Major Schofield, brother of a Civil War general, the Army orders three thousand Smith & Wesson .45 Schofield revolvers.

Since the barrel of the Smith and Wesson was shorter than that of the Colt, the .45 LC cartridge had to be adapted and the charge was reduced to 28 grains (1.81g). A few years later, the S&W

Schofield was withdrawn from service and sold in the private market.

Captain John R. Edie, officer in charge of the Ordnance Department for the evaluation of weapons reported: "I have no hesitation in declaring that the Colt revolver is superior in many ways and much more suited to the military than the. Smith and Wesson ".

The mechanism of Colt SAA may seem somewhat archaic; however, for a service weapon, it has the advantage of simplicity and strength because disassembly and reassembly are easy, dust and humidity are well tolerated. Certainly a part can break, but it will still be able to function. The strength of the Single Action is mainly due to its closed steel casing.

With the first eight thousand revolvers supplied between September 1873 and March 1874 to equip the ten cavalry regiments, the United States federal government ordered more than thirty-seven thousand until 1890.

The Colt SAA Cavalry Model played a large role in the Indian wars, it was, with the Springfield Model 1873 musket, the main weapon of the American cavalry. Every soldier and officer of the Seventh U.S. Cavalry Regiment under the command of General George Armstrong Custer wore a Colt SAA on their belt during the Battle of Little Big Horn against the Sioux and Cheyenne in 1876.

In 1887, Smith and Wesson produced the Smith & Wesson Safety Hammerless in cal. 32 and .38 until 1940, selling five hundred thousand copies. It is colloquially called the lemon squeezer. As its name suggests, this model of revolver has a safety device and no visible hammer (hammerless). These revolvers were very advanced because they have an inertia firing pin, developed by Joe Wesson, son of Daniel B. Wesson. It was mounted in the frame and spring loaded, it could strike the primer only when it received a sufficient blow from the hammer, transferring momentum to the firing pin itself. When the hidden hammer was at rest against the firing pin, it did not protrude from the frame.

In 1889, the army introduced a new type of revolver, in .38 caliber, with a falling barrel. The approximately fourteen thousand Colts SAA are removed and stored at the arsenal in Springfield, Massachusetts. Since the new revolver was not satisfactory because the stopping power was not sufficient, it was decided to restore the old Colts SAA in condition, to shorten the barrel to 5.5 inches and to return them to the troops. . These revolvers, said Artillery Model, will be delivered to troops in service during the Spanish-American War and the American-Philippine War.

At the age of 65, Smith retired and sold his share of the business to Wesson, making him the sole owner of the business. He died on January 15, 1893 and bequeathed all his property to works of public utility.

In 1899, Smith & Wesson introduced what is arguably the most famous revolver in the world, the .38 Military & Police (renamed Model # 10 in 1957). This revolver has been in continuous production from that year to the present day and has been used by virtually all police and military forces around the world.

Wesson remained active in the business until his death. He died at his home in Springfield, Massachusetts on August 4, 1906 after a long illness.

In 1901, shortly before Elizabeth Colt's death, the Colt family sold their gun business. Maintaining the Colt name, the new company continues to produce some of the best known and most reliable firearms (such as the infamous Colt .45) in American history.

The area in and around the factory complex rebuilt by Elizabeth Colt was listed on the National Register of Historic Places in June 1976 and remains a popular and memorable part of the Hartford landscape today.

ABOUT THE AUTHOR

Wandering philosopher, self-taught, conveyer of knowledge, very curious about the history of the world and lover of ancient know-how.

After having walked up and down the cadurque paths and having fallen completely under the spell of the causses and the endless forests of Quercy, he ended up installing his teepee in the heart of the Lotois woods (France), leading a healthy life mixed with a deep tranquility in the within these territories steeped in history.

Colt et Wesson, the story of a rivalry

www.ingramcontent.com/pod-product-compliance
Lightning Source LLC
Chambersburg PA
CBHW031551210526
45464CB00003B/1251